Army Service Commands of World War II

- *Their History and Insignia*

by

Michael D. Jorgensen

1st Edition — August 2011

ISBN: 978-0-9839125-0-7

Published by:

Michael D. Jorgensen
Post Office Box 142652
Fayetteville, GA 30214-6518

Dedication

This is the first of several works I anticipate publishing on military insignia. As it is my first work, I wish to dedicate it to my parents — Dale W. and Virginia A. Jorgensen of Hastings, Nebraska.

My parents encouraged each of their children to explore different things and pursue those which were of interest to each of them. While our interests have evolved and changed over the years, their support and love has contributed to each of us being able to realize many dreams and possibilities. My siblings and I are grateful for all that they have done and given.

Mike Jorgensen
August 2011

Glossary

ASMIC	The American Society of Military Insignia Collectors
DUI	Distinctive Unit Insignia (or unit crest)
FE	Fully Embroidered
GB	Greenback
NW	Northwest
OD	Olive Drab
O.Q.M.G.	Office of Quartermaster General
SC	Service Command
SSI	Shoulder Sleeve Insignia
TIOH	The United States Army Institute of Heraldry
U.S.	United States
W.A.A.C.	Women's Army Auxiliary Corps

Table of Contents

1st through 9th Service Commands

Formed during the Pre-World War II Emergency Period.

Inactivated June 11, 1946.

Northwest Service Command

Activated September 2, 1942 at White Horse (Yukon Territory), Canada.

Relocated to Edmonton (Alberta), Canada — February 1944.

Inactivated June 30, 1945.

Persian Gulf Service Command

Formed on August 11, 1942 when the Iran – Iraq Service Command
was re-designated at Basra, Iraq.

Relocated to Teheran, Iran — January 1943; subsequently relocated
to Khorramshahr, Iran — October 1945.

Inactivated December 31, 1945.

Introduction

Like many other collectors of Army shoulder sleeve insignia (SSI), one of the first pieces I acquired was that of a World War II era Army Service Command — the blue background and edged, fully embroidered, standard issue SSI. Completing the set along with a few variations was relatively easy and inexpensive. At some point the set of insignia became just another group of patches, relegated to the farthest back area of the collection. It was only when the pursuit of variations began did I realize there was much more to this group of insignia or SSIs than I had previously thought.

The purpose of this work is to provide a new insight into this perceived common group of insignia. Through it I hope that collectors of military insignia will not only become more aware of the history of these World War II era units but will also pursue these and other variations for their collection, especially those who are just beginning to collect. To assist collectors and others using this work, over 170 illustrations on twenty-four (24) color plates have been included.

Do other variations exist? The answer is unequivocally "*YES.*" How many others may exist? There is no definitive answer to this question. It may only be a small number or even a much larger quantity. I do believe that the insignia discussed in this work represent the major variations collectors will encounter.

In documenting the illustrations within this work, I have chosen to avoid the use of category numbers. Instead, I have organized illustrations by broad categories. Where appropriate, and in the case of minor variations, comments have been added to assist the user in distinguishing one insignia from another. I have also chosen not to include any information on the dimensions of insignia in an attempt to preclude or complicate their reproduction. The illustrations included generally approximate the size of the actual insignia but none are 100 percent images. Each illustrated insignia is an original World War II era piece.

Acknowledgements

Information used to develop this work was obtained from Department of the Army (The Institute of Heraldry) official files, *Shoulder Sleeve Insignia of the U.S. Armed Forces 1941 – 1945* by Richard W. Smith, and other information found in the public domain.

I especially wish to thank Dave and Steve Johnson (*a.k.a.,* the Johnson brothers), Bob Chatt, Chris Aleck and George Harrison for their contributions and assistance to this work. Special acknowledgement must also go to my wife Betty for her numerous ideas, assistance and unyielding support. *Thank you.*

Mike Jorgensen
August 2011

A History of World War II Army Service Commands

1st – 9th Corps Area Service Commands

With passage of the 1920 amendment to the National Defense Act of 1916, nine multi-state size Corps areas were established within the continental United States, by the United States Army Chief of Staff through War Department General Order 50, dated August 20, 1920.

The Corps areas were formed for the administration, training and tactical control of the United States Army, replacing the six geographical (or territorial) departments in which the continental United States had been divided since 1917 and with little variation since the Civil War. A Corps area consisted of divisions of the Regular Army, Army Reserve (Organized Reserves) and the National Guard of the United States.

The Corps Area Service Commands, as they were initially known, were formed during the pre-World War II emergency period. On October 3, 1940, the War Department transferred tactical command functions to General Headquarters, United States Army. A subsequent February 28, 1942 reorganization of both the Army and the War Department, placed the Corps Area Service commands under the War Department's *"Services of Supply."* The former Corps Headquarters were preparing for deployment to overseas theaters.

As part of another reorganization of the War Department on March 12, 1943, the *"Corps Area"* component of the designation was dropped with these organizations now being identified only as *"Service Commands."* Concurrently, the *"Services of Supply"* was re-designated as the *"Army Service Forces."*

The 1st – 9th Service Commands were inactivated on June 11, 1946 as a result of the War Department implementing a planned, post-war reorganization and downsizing. The nine Service Commands were replaced by six Army areas which roughly followed the old Corps area boundaries.

Service Command Mission

The Service Commands were responsible for non-combat related operations within the United States including: induction centers, administration, personnel record and mail service; supply operations, weapons issue and equipping of soldiers and units; operation of ports of embarkation; medical services; movement of supplies and troops to ports of embarkation; coordination of military ammunition production; providing instructors to conduct training; installation operation; development and distribution of educational programs, films and newspapers to soldiers; and other *"housekeeping"* functions in support of the war effort.

Table 1 provides a summary of the different Service Command headquarters location as well as their geographic areas of responsibility.

Table 1 – Army Service Command Locations and Geographic Areas

Army Service Command Headquarters Location and Geographic Areas of Responsibility		
Command	**Headquarters Location**	**Geographic Area of Responsibility**
1st Service Command	Boston, MA	Connecticut, Massachusetts, Maine, New Hampshire, Rhode Island and Vermont
2d Service Command	New York, NY (Fort Jay / Governors Island)	New Jersey and New York
3rd Service Command	Baltimore, MD (Fort McHenry)	Delaware, Maryland, Pennsylvania, Virginia and the District of Columbia
4th Service Command	Atlanta, GA (Fort McPherson)	Alabama, Florida, Georgia, Mississippi, North Carolina, South Carolina and Tennessee.
5th Service Command	Columbus, OH (Fort Hays)	Indiana, Kentucky, Ohio and West Virginia
6th Service Command	Chicago, IL (Fort Sheridan)	Illinois, Michigan and Wisconsin
7th Service Command	Omaha, NE (Fort Crook)	Colorado, Iowa, Kansas, Minnesota, Missouri, Nebraska, North Dakota, South Dakota and Wyoming
8th Service Command	San Antonio, TX (Fort Sam Houston) / Dallas, TX *	Arkansas, Louisiana, New Mexico, Oklahoma and Texas
9th Service Command	Presidio of San Francisco, CA / Fort Douglas, UT **	Arizona, California, Idaho, Montana, Nevada, Oregon, Utah and Washington

* Unit relocated from San Antonio to Dallas in 1942 occupying the Santa Fe Office Building.

** Unit relocated to Fort Douglas, UT in January 1942.

1st – 9th Service Command Insignia

Official War Department records reflect the design of the Service Commands insignia or SSI as having been an arbitrary decision. The initial color was an olive drab (OD) background insignia with a geometric pattern, representing the command's numeric designation, in white. White was selected for the geometric design due to it being a mixture of all colors and to reflect the Service Command being composed of all arms and services.

The change to a blue background insignia was directed by the following two pieces of War Department correspondence.

- Correspondence dated October 14, 1941. This correspondence does not indicate a specific reason for a change in the olive drab background to blue other than a remark of " *.... in conformity with informal request Assistant Chief of Staff, G-4.*" This same correspondence recommended that production awards not be made for olive drab background insignia.

- Correspondence from the War Department, Clothing Section C&E Branch to the Procurement Section, S&E Branch, dated October 16, 1941. It requested the "*Philadelphia Quartermaster Depot be contacted by telephone and advised to not make awards of the following quantities of Shoulder Sleeve Insignia, which is part of Procurement Directive P-C-180, which opened October 16, 1941 ....*" This request should have ceased efforts to produce olive drab background insignia.

Were OD background fully embroidered insignia produced? *YES*. It is unknown, however, which were produced and in what quantities. Actual production quantities are believed to be minimal.

Examples of two known OD background SSI are illustrated by *Plate 1* – insignia of the 4th and 5th Service Commands. Do other examples of original fully embroidered OD background SSIs exist? They may, but the author has no knowledge of them. However, the author is aware of the existence of a "*felt on wool*" 1st Service Command "*oval.*" It is unknown if this insignia is an original or a reproduction piece. Due to this uncertainty, this insignia has not been included within this work.

A summary of United States War Department, Office of Quartermaster General (O.Q.M.G.) approval dates for each Service Command SSI is at *Table 2*. As noted by this table, the "*oval*" 1st Service Command had a very short duration. This short time period may explain the rarity of these pieces. A search of official government files could not locate any explanation for the change in design.

Illustrations of the 1st Service Command "*oval*" with "*olive drab*" and "*blue*" edges are shown at *Plate 2*. Collectors will find each design has been reproduced.

In the last few years, collectors have also been offered branch color variations of the "*oval*" shaped 1st Service Command SSI. Information accompanying these offerings is intentionally vague, in the opinion of the author, with the insignia being offered at inflated or unrealistic prices. Insignia as these should fall into the category of "*novelty*" or "*fantasy*" pieces.

Table 2 – Service Command Insignia Approval Dates

Unit / Organization	United States War Department (O.Q.M.G.) Approval Dates	
	Olive Drab (OD) Background	Blue Background
1st Service Command – *Oval*	July 17, 1941	October 14, 1941
1st Service Command – *Square*	N/A	November 28, 1941
2d Service Command	July 18, 1941	October 14, 1941
3rd Service Command	July 17, 1941	October 14, 1941
4th Service Command	July 18, 1941	October 14, 1941
5th Service Command	July 18, 1941	October 14, 1941
6th Service Command	July 17, 1941	October 14, 1941
7th Service Command	July 18, 1941	October 14, 1941
7th Service Command – *Alternative Design (Unofficial)*	N/A	N/A
8th Service Command	July 18, 1941	October 14, 1941
9th Service Command	July 17, 1941	October 14, 1941

As previously stated, and as noted by the information in *Table 2* above, the "*oval*" shaped 1st Service Command insignia design existed for just over a month before it was replaced. The potential of "*colored*" branch variations of this SSI being officially authorized by O.Q.M.G. does not pass a reasonable logic test, especially when one considers this timeline and conditions of the era. Additionally, the command had only begun its mission. Required resources to design and produce branch colored insignia would have also been in short supply. These points and others support the argument colored branch insignia were not officially produced. Other priorities would have clearly existed.

A point of trivia. Among its many responsibilities, O.Q.M.G., had the responsibility of designing and authorizing insignia for both the United States Army and the United States Army Air Forces. O.Q.M.G. was the forerunner of today's United States Army "*The Institute of Heraldry.*"

Service Command Insignia Variations

Collectors will find that the most common variations of Service Command insignia are those having an embroidered blue background with a white geometric design. While variations exist in the embroidery of the background material weave or pattern, most variations are due to embroidery of the geometric design — both in its detail and pattern. These variations are the primary focus of this work.

The following provides a summary of the insignia variations which are known to exist for the 1st through 9th Service Commands.

Olive Drab (OD) Border, Fully Embroidered. These are the earliest examples of the 1st through 9th Service Command SSI. While it is unknown how long OD border SSIs were produced, it is believed they were the first production variation with the blue border becoming the more commonly known insignia. This change most likely occurred soon after the beginning of World War II.

The background material and geometric designs for the OD border SSIs are both fully embroidered. *Plate 3* illustrates known 1st through 9th Service Command OD border insignia. Please note the variations of the geometric design for the insignia of the 9th Service Command. It is unknown if other variations in the geometric design exist.

The rarity of the OD border pieces varies considerably. Some are relatively common and inexpensive; others are rare, drawing high values from collectors.

Twill background. Twill background varieties, with embroidered geometric designs, also exist for the 1st through the 9th Service Commands (see *Plate 5*).

An unusual variation is that of an OD border, twill background insignia of the 6th Service Command (see *Plate 4*). The author has no information if similar insignia may exist for other commands.

Embroidery variations. As previously stated, if one makes even a *"shallow dive"* into the standard issue fully embroidered SSI, the large number of variations is quickly apparent. Again, most variations will exist in the embroidery of the geometric design. Known variations are illustrated by *Plate 6* through *Plate 16*.

Variations range from the "double wide border" of the 1st and 3rd Service Command, to the differences in the embroidery of 2d Service Command interlocking squares. These and the many other embroidery variations offer collectors many opportunities.

Some insignia can be oriented in different ways and still be correctly displayed. The orientation will affect the direction of the weave and cause a vertical, horizontal or diagonal pattern to appear. Such differences do not constitute a variation.

The size of a geometric design, concurrent with the design having a *"finished"* inner or outer edge, creates other variations. The color and style of the outer and/or inner edge of the geometric design can also be different. Some designs have an unfinished edge(s), others have a white edge and yet others a blue edge or a combination thereof.

The reverse embroidery of these insignia is normally white in color but they can also be found with a green color. These differences are due to the color of the *"return"* or *"catch"* thread used during the manufacturing process. Variations such as this are known as *"greenbacks"* or *"snowbacks"* respectively. *Greenback* insignia are known to exist for all commands. Illustrations are included for all of these insignia except the 6th Service Command.

7th Service Command – Alternative Design

Taking a step back, it is time to review what is frequently referred to as the *"Officer"* design of the 7th Service Command SSI. Documentation confirming this insignia as having been authorized by the Office of the Quartermaster General (O.Q.M.G.) cannot be located in official government files or other existing references - even though the use of this insignia is known to have occurred and is widely accepted.

Available references, including the *ASMIC Trading Post* of October – December 1971, reflect these insignia design being worn by both officers and enlisted personnel. These references do not explain why this design was created. Due to the absence of official documentation, the author has chosen to call this an *"alternative design."* Please see *Table 2* for information regarding insignia approval dates.

Alternative design insignia are known to exist in all materials except that having an OD border. *Plate 17* provides an illustration of known variations. Among these variations is a unique and beautiful "beaded" insignia. A unit number, applied to an SSI, adds to the list. It is unknown which unit this number represents. It may be that of a 7th Service Command Army Hospital. The author believes the "pink" outer and inner border of the insignia illustrated on *Plate 17* was originally "red." Use and laundering have caused the color to fade.

Other Insignia Variations

The insignia previously discussed are complemented by the existence of those having a felt background. The geometric designs can be found in felt, full embroidery and bullion embroidery. Examples of several variations are at *Plate 18* and *Plate 19*. Variations of other bullion embroidery on felt insignia are known to exist.

DUI size. Small insignia, approximately the size of a Distinctive Unit Insignia (DUI) or a *"unit crest,"* exist in two different color variations – white embroidery on blue felt and blue embroidery on white twill. *Plate 20* provides illustrations of known insignia including a design unique to the 7th Service Command at Camp Crowder, MO. These insignia are believed to have been worn on an enlisted soldiers service hat.

A number of other variations of 1st through 9th Service Command insignia exist. *Plate 21* illustrates many of those known including:

- **Geometric design rotated center**. A variety that is frequently overlooked by most collectors is that of the 5th Service Command having a *"rotated"* center.

- **Reverse or other color background material or geometric designs**.

  Reasons for the existence of reverse or other color insignia have yet to be located. It is the opinion of some that colored insignia were created to reflect the branch designation of military instructors or the type of training being conducted (e.g., Infantry, Artillery, etc.). Unfortunately, the author has been unable to locate any evidence to support these opinions. To reinforce a statement in the *Introduction* to this work, each insignia illustrated are World War II era originals.

- **Branch color trimmings**. The application of colored branch trimmings around the outer edge of a Service Command insignia also occurred. This was somewhat of a common practice by soldiers during World War II.

  Plate 21-1 provides an illustration of the use of Medical and Military Police branch trimmings applied to two different 4th Service Command SSIs.

- **Other insignia variations**.

  Variations of World War II era Service Command insignia also include the use of installation names or specific functions embroidered on an insignia. Among the illustrations on *Plate 21* is a superb 9th Service Command Military Police insignia. Information on how this insignia was used has yet to be located. Possible uses include it being used as a pocket patch or on a brassard by military or civilian police.

  The name of Camp Crowder, a 7th Service Command installation located near Neosho, Missouri, appears on a large number of different insignia, including an OD border SSI. Other variations also include a shoulder arc applicable to Camp Crowder and an 8th Service Command Post Engineer installation activity.

- **Women's Army Auxiliary Corps (W.A.A.C.) Recruiting Brassards**.

  A number of these felt brassards exist. It is unknown, however, if each command or state within a command had a designated brassard. Brassards were created by an insignia being directly embroidered on it. The insignia is larger than the DUI size insignia previously discussed but much smaller than a standard issue SSI. An example of one brassard is shown on *Plate 22*.

  Also illustrated on *Plate 22* is a fully embroidered 8th Service Command insignia on a yellow felt background. Upon examination, this insignia is found to be the same size as that embroidered on a brassard. The author believes this insignia was "*cut out*" or removed from a brassard.

  The piece described was offered to a collector with the "*annotation*" it was used as a 8th Service Command "*Cavalry*" hat insignia. While this statement may be true, with a serviceman having cut the insignia from a brassard, a possibility exists that this insignia reflects an effort by someone to promote an insignia for other than what it may actually be.

Northwest Service Command

The Northwest Service Command was activated on September 2, 1942. The SSI of this command was approved on March 23, 1943.

The headquarters of the Northwest Service Command was initially located at White Horse (Yukon Territory), Canada. The command relocated to Edmonton (Alberta), Canada in February 1944.

The primary mission of the command was the construction of the Alaska-Canada (ALCAN) Highway and, subsequently, its operation and maintenance. The command had the same power and authority over U.S. Army activities in Western Canada as did the 1st through 9th Service Commands located in the continental United States.

The Northwest Service Command was inactivated on June 30, 1945. The 6th Service Command assumed the command's functions, transferring the Canadian portion of the ALCAN Highway to the Canadian government in 1946.

As with the insignia of the 1st through 9th Area Service Commands, variations of this command's insignia exist. *Plate 23* illustrates the standard issue, fully embroidered, insignia of the command as well as theater made examples.

Please note the difference between the two standard issue insignia. The blue field of the insignia illustrated on the right has a finished inner blue border where the other does not. While minor, these differences do constitute a variation.

Two examples of theater made insignia are also illustrated - one embroidered on wool and another handmade from ribbon material. Other theater made examples are known to exist.

Persian Gulf Service Command

Depending upon one's point of view, it can be argued this organization was a "*Service Command*" or a "*Theater Operational Command.*" Due to the unit being designated as a Service Command for two different periods, these insignia have been included within the scope of this work. It is recognized that the mission of the Persian Gulf Service Command was different from that of the other Service Commands.

The Persian Gulf Service Command was formed on August 11, 1942 when the Iran – Iraq Service Command was re-designated. The unit's headquarters was initially at Basra, Iraq; subsequently relocating to Teheran, Iran in January 1943.

On December 10, 1943, the Persian Gulf Service Command was re-designated as the Persian Gulf Command. At the same time, the Persian Gulf was designated as a separate theater of operations.

The unit was again re-designated as the Persian Gulf Service Command in October 1945 with the unit's headquarters being moved to Khorramshahr, Iran.

The unit's SSI was approved on April 29, 1944 and may have been initially classified.

The mission of the Persian Gulf Service Command was to ensure the uninterrupted supply of war-material being sent to Russia (Soviet Union) under the United States lend-lease program. Following the end of World War II, the mission of the command was to close all United States Army installations and activities in the Persian Gulf area.

The command was inactivated December 31, 1945.

As with other insignia addressed in this publication, the insignia of the Persian Gulf Service Command exist as a standard issue, fully embroidered, SSI as well as those which are theater made. An illustration of a standard issue insignia having bullion applied embroidery is included (see *Plate 24*) as well as others which are German made. Several superb examples of known theater-made SSI have also been included. Among these is a unique felt background shoulder arc with bullion embroidery as well as a beautiful "*inverted*" insignia. This insignia is of special interest as it brings to question how was it to be worn on the uniform — pointed end up or pointed end down? Please see *Plate 24-2*.

Summary

Where does the list of World War II era Service Command insignia variations end? Why do so many different variations exist? Did standards not exist for the design and production of military insignia during World War II?

A complete answer to each question may never be known. What can be said though is that the insignia of these commands are among the most abundant of the World War II era. The thousands of soldiers assigned to the different commands would have required significant quantities of insignia.

To meet the demands for insignia over the length of time these units existed, different manufacturers were undoubtly contracted by the War Department. It is also known that over the course of World War II, the techniques and materials used to produce insignia changed - most likely due to the types of material available. Collectively, these factors created the *"perfect storm,"* subsequently resulting in the large number of different variations illustrated within this work.

The insignia discussed by this work present a unique opportunity and, in some cases, challenges for the collector of military insignia. While a collector can make a decision at any point regarding what to collect, the "door can be left open" to support collecting any number of variations. Many of the variations illustrated can be found in what some refer to as "junk boxes" at military collector shows, Army — Navy stores, flea markets, antique shops, vendors specializing in collectables, etc..

While other considerations do exist, a collector or historian can use the insignia of World War II Service Commands as a baseline to expand their study of other insignia. As an example, a number of variations exist for the insignia of Army Divisions, other World War II era commands and even the insignia of modern or post-World War II commands.

In summary, I trust this work will positively assist collectors, historians and others in multiple ways and that it will serve as a valuable reference.

This Page Left Blank for Notes

PLATE 1
Initial Design, Olive Drab Background (FE)

☐ **4th Service Command**

☐ **5th Service Command**

PLATE 2
1st Service Command, Oval Design (FE)

Obverse

Reverse

❐ **1st Service Command
OD Border**

❐ **1st Service Command
Blue Border "Greenback"**

PLATE 3
1st – 9th Service Command, Olive Drab Border (FE)

❏ **1st Service Command**

❏ **2d Service Command**

❏ **3d Service Command**

❏ **4th Service Command**

❏ **5th Service Command**

❏ **6th Service Command**

❏ **7th Service Command**

❏ **8th Service Command**

❏ **9th Service Command**
   (Circular pattern)

❏ **9th Service Command**
   (Diagonal pattern)

PLATE 4
1st – 9th Service Command, OD Border - Twill Background

☐ 6th Service Command

PLATE 5
1ˢᵗ – 9ᵗʰ Service Command, Twill Background

☐ **1st Service Command**

☐ **2d Service Command**

☐ **3d Service Command**

☐ **4th Service Command**

☐ **5th Service Command**

☐ **6th Service Command**

☐ **7th Service Command**

☐ **8th Service Command**

☐ **9th Service Command**

PLATE 6 — 1
1st Service Command (FE)

☐ **1st Service Command**

Outside and inside edges of the
design are <u>finished</u> in <u>white</u>.
The "one" is <u>wide</u>. A
horizontal pattern in the design.

☐ **1st Service Command**

Outside and inside edges of the
design are <u>finished</u> in <u>white</u>.
The "one" is <u>narrow</u>. A
vertical pattern in the design.

☐ **1st Service Command
"Greenback"**

Outside and inside edges
of the design are <u>unfinished</u>.
The "one" is <u>narrow</u>. A
vertical pattern in the design.

PLATE 6 — 2
1st Service Command (FE)

❏ **1st Service Command**

Outside and inside edges of the
design are <u>finished</u> in <u>blue</u>.
The "one" is <u>wide</u>. A
vertical pattern in the design.

❏ **1st Service Command**

Outside and inside edges of
the design are <u>unfinished</u>.
The "one" is <u>narrow</u>. A
vertical pattern in the design.

❏ **1st Service Command**

Outside edges of the design are
<u>finished</u> in <u>blue</u>. Top and bottom
of the "one" are also <u>finished</u>
in <u>blue</u>; sides are <u>unfinished</u>.
The "one" is <u>wide</u>. A vertical
pattern in the design.

19

PLATE 7 — 1
1st Service Command, Double Wide Border (FE)

❐ 1st Service Command

Outside and inside edges of
the design are <u>unfinished</u>. The
"one" is <u>narrow</u>. A vertical
pattern in the design.

❐ 1st Service Command

Outside and inside edges of
the design are <u>finished</u> in
<u>white</u>. The "one" is <u>wide</u>. A
horizontal pattern in the design.

❐ 1st Service Command

Outside and inside edges of the
design are <u>unfinished</u>. The
"one" is <u>wide</u>. A vertical
pattern in the design.

❐ 1st Service Command

Outside edges of the design are
<u>unfinished</u>. The "one" is <u>narrow</u>
and <u>embroidered</u> on the design;
edges are <u>unfinished</u>. A
diagonal pattern in the design.

PLATE 7 — 2
1st Service Command, Double Wide Border (FE)

☐ **1st Service Command
"Greenback"**

Outside edge of the design is
<u>unfinished</u>. The "one" is <u>wide</u>
and embroidered on the design;
edges are <u>finished</u> in <u>blue</u>. A
diagonal pattern in the design.

☐ **1st Service Command
"Greenback"**

Outside and inside edges of the
design are <u>unfinished</u>. The "one" is
<u>narrow</u> . A vertical pattern in the design.

PLATE 8 — 1
2d Service Command (FE)

Embroidery of the geometric design is offset from edge of the insignia.

❐ **2d Service Command**

Outside and inside edges of the design
are <u>unfinished</u>; inside is <u>narrow</u>. A
horizontal pattern in the design.

❐ **2d Service Command**

Outside and inside edges of the design
are <u>unfinished</u>; inside is <u>narrow</u>. A
horizontal pattern in the design. The
design is smaller than others illustrated.

❐ **2d Service Command**

Outside and inside edges of the design are
<u>unfinished</u>; inside is <u>wide</u>. The corners of each
design segment have an embroidered line.

❐ **2d Service Command**

Outside and inside edges of the design
are <u>finished</u> in <u>white</u>; inside is <u>narrow</u>.
A horizontal pattern in the design.

❐ **2d Service Command
"Greenback"**

Outside and inside edges of
the design are <u>unfinished</u>;
Inside is <u>narrow</u>. A minor
pattern in the design.

22

PLATE 8 — 2
2d Service Command (FE)

Embroidery of the geometric design is at edge of the insignia.

☐ **2d Service Command**

Outside and inside edges of the
design are <u>unfinished</u>; inside
area is <u>wide</u>. A vertical pattern
in the design.

☐ **2d Service Command**

Outside and inside edges of the
design are <u>unfinished</u>; inside is
<u>narrow</u>. A vertical pattern
in the design.

☐ **2d Service Command**

Outside and inside edges of
the design are <u>finished</u> in
<u>white</u>; inside area is <u>narrow</u>. A
horizontal pattern in the design.

☐ **2d Service Command**

Outside and inside edges of
the design are <u>finished</u> in <u>white</u>;
inside is <u>narrow</u>. A vertical pattern is
visible in the design. The design is
<u>larger</u> than the others illustrated.

☐ **2d Service Command**

Outside and inside edges of
the design are <u>finished</u> in <u>white</u>;
inside is <u>narrow</u>. No distinct
design pattern. The design is
<u>larger</u> than the others illustrated.

PLATE 9
3rd Service Command (FE)

❐ **3d Service Command**

Outside and inside edges of the design are <u>unfinished</u>. A distinct ribbed pattern is visible in the design.

❐ **3d Service Command**

Outside and inside edges of the design are <u>unfinished</u>. A minor pattern is visible in the design.

❐ **3d Service Command**

Outside and inside edges of the design are <u>finished</u> in <u>white</u>. A visible pattern in the design.

PLATE 10
3rd Service Command, Double Wide Border (FE)

☐ **3d Service Command**

Outside and inside edges of the design are <u>unfinished</u>. A visible pattern in the design.

☐ **3d Service Command**

Outside and inside edges of the design are <u>finished</u> in <u>white</u>. A visible pattern in the design.

☐ **3d Service Command**

Outside and inside edges of the design are <u>finished</u> in <u>blue</u>. A minor pattern in the design.

☐ **3d Service Command**
     "Greenback"

Outside and inside edges of the design are <u>unfinished</u>. A visible pattern in the design.

PLATE 11 — 1
4th Service Command (FE)

❏ **4th Service Command**

Arcs and ends of the design segments
have an elaborate embroidery.

❏ **4th Service Command**

A circular pattern in the
embroidery of the design.

❏ **4th Service Command**

Outside and inside edges of the design are
<u>unfinished</u>. The design has a visible pattern.

❏ **4th Service Command**
"Greenback"

Outside and inside edges of the
design are <u>unfinished</u>. A minor pattern
is visible in the design.

PLATE 11 — 2
4th Service Command (FE)

☐ **4th Service Command**   ☐ **4th Service Command**   ☐ **4th Service Command**

Outside and inside edges of the insignia shown are <u>finished</u> in <u>white</u>. The
difference between the three (3) images illustrated is the width of the center design.
An approximate 1/8" difference exists between the smallest and largest dimension
of the center in the three (3) insignia illustrated.

☐ **4th Service Command**
"Greenback"

Outside and inside edges of the design are <u>finished</u>
in <u>white</u>. The design has a visible pattern.

PLATE 12
5th Service Command (FE)

□ **5th Service Command**

Outside and inside edges of the design are <u>finished</u> in <u>blue.</u>

□ **5th Service Command**

Outside and inside edges of the design are <u>unfinished</u>. Insignia is 1/8" larger than the others Illustrated. WWII era vendor price tag attached.

□ **5th Service Command**

Outside and inside edges of the design are <u>finished</u> in <u>blue</u>. Inside area of the design is larger than the other illustrated.

□ **5th Service Command**

Outside and inside edges of the design are <u>finished</u> in <u>white</u>.

□ **5th Service Command "Greenback"**

Outside and inside edges of the design are <u>finished</u> in <u>white</u>.

□ **5th Service Command**

Outside and inside edges of the design are <u>unfinished</u>. The design has a circular pattern.

PLATE 13
6th Service Command (FE)

☐ **6th Service Command**

Outside and inside edges of the design are <u>unfinished</u>. Design segments are separated by an embroidered line. Design is <u>offset</u> from the border of the insignia.

☐ **6th Service Command**

Outside and inside edges of the design are <u>unfinished</u>. No line separates design segments. Design is <u>offset</u> from the border of the insignia.

☐ **6th Service Command**

Outside and inside edges of the design are <u>unfinished</u>. No line separates design segments. Design is <u>at</u> the border of the insignia.

☐ **6th Service Command**

Outside and inside edges of the design are <u>finished</u> in <u>white</u>. No line separates design segments. Design is <u>at</u> the border of the insignia.

PLATE 14 — 1
7th Service Command (FE)

❑ **7th Service Command**

Outside edge of the design is <u>unfinished</u>; inside edge is <u>finished</u> in <u>white</u>. The design has a circular pattern.

❑ **7th Service Command**

Outside and inside edges of the design are <u>unfinished</u>. The design has a circular pattern. The center of the design is larger.

❑ **7th Service Command**

Outside and inside edges of the design are <u>unfinished</u>. The design has a circular pattern.

❑ **7th Service Command**

Outside and inside edges of the design are <u>unfinished</u>. The design has a visible pattern.

❑ **7th Service Command
"Greenback"**

Outside and inside edges of the design are <u>unfinished</u>. The design has no distinct pattern.

❑ **7th Service Command**

Outside and inside edges of the design are <u>finished</u> in <u>white</u>. The design has a visible pattern.

30

PLATE 14 — 2
7th Service Command (FE)

☐ **7th Service Command**

**Outside edge of the design is
<u>unfinished</u>. The center of the
design has elaborate embroidery.**

☐ **7th Service Command
"Greenback"**

**Outside edge of the design is
<u>unfinished</u>. The center of the
design has elaborate embroidery.**

PLATE 15
8th Service Command (FE)

8th Service Command

Outside edge of the design is <u>unfinished</u>; inside edge is <u>finished</u> in <u>white</u>. The design has a circular pattern.

8th Service Command

Outside and inside edges of the design are <u>finished</u> in <u>white</u>. The design segments have a visible pattern. The center is <u>larger</u> than the others illustrated.

8th Service Command

Outside edge of the design is <u>unfinished</u>; inside edge of the design is <u>finished</u> in <u>white</u>. The design segments have a visible pattern.

8th Service Command

Outside and inside edges of the design are <u>finished</u> in <u>white</u>. Design segments have a visible pattern.

8th Service Command

Outside and inside edges of the design are <u>finished</u> in <u>blue</u>. Design segments have a visible pattern.

8th Service Command "Greenback"

Outside and inside edges of the design are <u>unfinished</u>. Design segments have a minor pattern.

PLATE 16
9th Service Command (FE)

☐ **9th Service Command**

Outside and inside edges of the design are <u>unfinished</u>. The design has a visible pattern.

☐ **9th Service Command**

Outside and inside edges of the design are <u>unfinished</u>. The design has a circular pattern.

☐ **9th Service Command**
"Greenback"

Outside and inside edges of the design are <u>unfinished</u>. The design has a visible pattern.

☐ **9th Service Command**

Outside and inside edges of the design are <u>finished</u> in <u>white</u>. The center of the design is <u>larger</u> than the others illustrated. The design is <u>at</u> the border of the insignia.

☐ **9th Service Command**

Outside and inside edges of the design are <u>finished</u> in <u>white</u>. The design is <u>at</u> the border of the insignia.

33

PLATE 17 — 1
7th Service Command, Alternate Design

❒ **7th Service Command**

White on blue, FE.

❒ **7th Service Command**

White on blue,
embroidery on twill.

❒ **7th Service Command**

White on black,
embroidery on twill.

PLATE 17 — 2
7th Service Command, Alternate Design

□ **7th Service Command**

White embroidery applied to a black felt background.
The insignia illustrated at right is larger.

□ **7th Service Command**

□ **7th Service Command**

White felt on black felt.

□ **7th Service Command**

Silver bullion embroidery on a
black felt background.

□ **7th Service Command**

Blue and white beads embroidered on
an insignia; dog tag chain border.

□ **7th Service Command**

White felt on blue felt with
the design outlined in <u>red</u>.

□ **7th Service Command**

An embroidered design and number "2"
applied to black felt background.

PLATE 18
1st – 9th Service Command

☐ **1st Service Command**

White felt on blue felt. "One" is from the background material.

☐ **1st Service Command**

Blue areas are embroidered on white felt.

☐ **1st Service Command**

Blue felt on white felt.

☐ **2d Service Command**

Silk screen printing on blue felt. Green felt background.

☐ **3d Service Command**

Embroidery on blue felt. Outside and inside edges of the design are <u>finished</u> in <u>white</u>.

☐ **6th Service Command**

White areas are embroidered on blue felt. Design segments are separated.

☐ **6th Service Command**

Pieced construction, white cord embroidery applied to edges. Blue felt background.

☐ **7th Service Command**

White areas are embroidered on blue felt. Outside and inside edges of the design are <u>finished</u> in <u>white</u>.

☐ **9th Service Command**

White felt on blue felt.

PLATE 19
1st – 9th Service Command, Bullion Embroidered

❏ **2d Service Command**

Silver bullion embroidery on a
blue felt background.

❏ **8th Service Command**

Post WWII German made insignia.
German aluminum embroidery
on a blue felt background.

❏ **9th Service Command**

Silver bullion embroidery on a
blue felt background.

PLATE 20
1ˢᵗ – 9ᵗʰ Service Command, DUI Size Insignia

Embroidered on Felt

❒ **1st SC**   ❒ **2d SC**   ❒ **3d SC**   ❒ **4th SC**   ❒ **5th SC**

❒ **6th SC**   ❒ **7th SC**   ❒ **8th SC**   ❒ **9th SC**

Embroidered on Twill

❒ **4th SC**   ❒ **5th SC**   ❒ **6th SC**

❒ **7th SC**   ❒ **9th SC**

❒ **3d Service Command**
Fully embroidered.

❒ **7th Service Command**
Embroidered on felt.

PLATE 21 — 1
1st – 9th Service Command, Other Variations

☐ **2d Service Command**

Outside and inside edges of the design are <u>finished</u> in <u>white</u>. White on black embroidery.

☐ **4th Service Command**

Outside and inside edges of the design are <u>finished</u> in <u>white</u>. Medical Corps branch piping applied.

☐ **4th Service Command**

Outside and inside edges of the design are <u>finished</u> in <u>white</u>. Military Police Corps branch piping applied.

☐ **5th Service Command**

Rotated center, FE.

☐ **5th Service Command**

Reverse colors, FE.

☐ **7th Service Command**

White on blue, FE.

☐ **7th Service Command**

OD border with embroidered lettering (FE).

PLATE 21 — 2
1ˢᵗ – 9ᵗʰ Service Command, Other Variations

☐ 8th Service Command

Outside and inside edges of the design are <u>finished</u> in <u>white</u>. White on black embroidery.

☐ 8th Service Command

Embroidery on twill.

☐ 9th Service Command

Embroidery on felt.

☐ 9th Service Command

Outside and inside edges of the design are <u>finished</u> in <u>red.</u> Blue on red embroidery.

☐ 9th Service Command

Outside and inside edges of the design are <u>finished</u> in <u>white</u>. Red on white embroidery.

PLATE 22
1st – 9th Service Commands, Recruiting Brassards

□ **4th Service Command**

W.A.A.C. Recruiting Brassard -
State of Georgia.

□ **8th Service Command**

Embroidery on felt.
Possible "Cavalry" hat insignia.

Comparison illustrations of
brassard insignia and separate
8th Service Command
"Cavalry" hat insignia.

PLATE 23
Northwest Service Command

❏ **Northwest Service Command**

Standard World War II issue (FE).
An embroidered <u>white</u> line borders
each side of the center field. Blue
and red fields are separated by an
embroidered <u>red</u> line.

❏ **Northwest Service Command**

Standard World War II issue (FE).
Lines separating the different
color fields are unfinished.

❏ **Northwest Service Command**

Theater made using
ribbon material.

❏ **Northwest Service Command**

Theater made using embroidery
applied to a wool background.

PLATE 24 — 1
Persian Gulf Service Command

☐ **Persian Gulf Service Command**

Standard World War II issue (FE).

☐ **Persian Gulf Service Command**

Standard World War II issue (FE) with bullion embroidery applied to design segments.

☐ **Persian Gulf Service Command**

Bullion embroidery applied to an embroidered background.

☐ **Persian Gulf Service Command**

☐ **Persian Gulf Service Command**

Post World War II German made insignia (FE).

☐ **Persian Gulf Service Command**

Bullion embroidery applied to an embroidered background.

PLATE 24 — 2
Persian Gulf Service Command

❏ **Persian Gulf Service Command**

❏ **Persian Gulf Service Command**

Embroidery on felt.

❏ **Persian Gulf Service Command**

Embroidery applied to a suede background. Bullion borders and highlights.

❏ **Persian Gulf Service Command**

Embroidery applied to a felt background. Bullion borders and highlights. Inverted design.

❏ **Persian Gulf Service Command Shoulder Tab**

Post World War II insignia — theater made (bullion embroidery on felt).

❏ **Persian Gulf Service Command**

Embroidery applied to an embroidered weave background.

This Page Left Blank for Notes

This Page Left Blank for Notes

www.ingramcontent.com/pod-product-compliance
Lightning Source LLC
Chambersburg PA
CBHW061049090426
42740CB00002B/92